YOUR KNOWLEDGE HAS VALUE

- We will publish your bachelor's and master's thesis, essays and papers

- Your own eBook and book - sold worldwide in all relevant shops

- Earn money with each sale

Upload your text at www.GRIN.com
and publish for free

Bibliographic information published by the German National Library:

The German National Library lists this publication in the National Bibliography; detailed bibliographic data are available on the Internet at http://dnb.dnb.de .

This book is copyright material and must not be copied, reproduced, transferred, distributed, leased, licensed or publicly performed or used in any way except as specifically permitted in writing by the publishers, as allowed under the terms and conditions under which it was purchased or as strictly permitted by applicable copyright law. Any unauthorized distribution or use of this text may be a direct infringement of the author s and publisher s rights and those responsible may be liable in law accordingly.

Imprint:

Copyright © 2018 GRIN Verlag
Print and binding: Books on Demand GmbH, Norderstedt Germany
ISBN: 9783668704039

This book at GRIN:

https://www.grin.com/document/424947

Huraiza Ramli

Determination of Alcohol Content in Halal Beverages

GRIN Verlag

GRIN - Your knowledge has value

Since its foundation in 1998, GRIN has specialized in publishing academic texts by students, college teachers and other academics as e-book and printed book. The website www.grin.com is an ideal platform for presenting term papers, final papers, scientific essays, dissertations and specialist books.

Visit us on the internet:

http://www.grin.com/

http://www.facebook.com/grincom

http://www.twitter.com/grin_com

DETERMINATION OF ALCOHOL CONTENT IN HALAL BEVERAGES

Huraiza Ramli

Processing and Quality Control of Food Unit, Kolej Komuniti Sik

ABSTRACT

The study involved the detection of the alcohol content of various market halal beverages products. The objectives of this study are to identify the food products that contain alcohol and to quantify the percentage of ethanol present. Gas Chromatography-Mass Spectrometer (GC-MS) was used to screen the samples for detection of ethanol and further quantification. Samples were purchased from several hypermarkets in Kuantan. A total of 95 samples, including fermented and cultured beverages, carbonated drinks, juices and cordials, tea and coffee, energy drinks, colouring and flavourings, and vinegars were analyzed. Of the total number of samples screened, 58 were found to contain ethanol. The ethanol content in the samples ranged from 0.001% to 3.510%. The GC-MS method is proven to be an effective and reliable method for the detection of low alcohol concentrations by minimizing the loss of the ethanol compound during sample preparation and extraction. Ethanol is present due to natural fermentation of the product or added as solvent during production.

Keywords: alcohol content, halal food, GC-MS

1. INTRODUCTION

Foods are the basic needs for human being survivor and also play an important role in a community's religious, social and culture. Islam is the world's second largest religion with the estimation population of 1.3 billion worldwide (Riaz and Chaudry, 2004). Awareness of Muslim consumers toward the validity of the halal status and prohibition in choosing the food product. Issues had been raised about the validity of the halal status of a certain market food products.

Alcohol is widely used in the food industry as solvents in flavours, colours and preservatives. Besides foods, alcohols are also widely used in cosmetics, pharmaceuticals and other industrial products. The total value of alcohol used in these industries amounts to billions of dollars and in the soft drink sector alone, Coca Cola reported sales of USD22 billion in its 2004 annual report (Azahari, 2010). Based on the amount of alcohol used in the food and other industries, it is difficult for any food industry to switch their industrial practices from using alcohol to alternative ingredients (Law *et al.*, 2011). Currently, some of the issues related to the halalness of food products are the presence of alcohol in foods and beverages (Riaz, 1997). One of the critical parts in issuing halal certificates is the level of permissible alcohol content in food products.

Among all, determination of the alcohol, especially ethanol content of in food products become one of the critical issue. Ethanol is a common chemical in nature, produced during fermentation or added for the aid of processing. Alcohol, especially ethanol consumption has health and social consequences via intoxication (drunkenness), alcohol dependence, and other biochemical effects of alcohol (WHO, 2004). Ethanol acts as a drug affecting the central nervous system. Its behavioral effects stem from its effects on the brain. Consumption of alcohol in any form is totally prohibited in the Quran and for the Muslim community. The Arabic term used for alcohol in the Quran is "khamr", implies not only to alcoholic beverages but also to all things that intoxicate or affect one's thought process.

Ethanol compound has several unique properties that give rise to the difficulty in their analytical and interpretive procedures. Gas chromatography had been widely used in determination of the alcohol in foods and beverages either for scientific research or routine analysis in the industry (GonzaÂlez-Arjona et al., 1999; Lachenmeier et al.,2008) Headspace sampling has shown to be a simple, rapid, solvent less, and reliable technique. Study aim to identify the food products that contained ethanol and to quantify the percentage of ethanol present. Data can be collected and used as part of data based for alcohol content of market beverages in Malaysia.

1.1 OBJECTIVES

The objectives of this research are:
1. To identify the food products that contained ethanol.
2. To quantify the percentage ethanol content in food products by using GC-MS.

1.2 PROBLEM STATEMENTS

The problem statements are :
1. Issues had been raised about the validity of the halal status of some certain market food products.
2. Many issues on doubtful of consuming halal food that contain alcohol.

2. LITERATURE REVIEW

A controversial issue on grey areas of Halal is the presence of alcohol in food and beverage. Traditionally, consumers and Islamic jurists have identified alcohol as a substance that is Haram for consumption, whilst the process of fermentation is perceived as an unethical process as it produces intoxicants.

Since alcohol exists in small quantities in Halal food products, consumers are unsure of its legal values and whether it can be consumed. However, food producer claims that the fermentation processes itself are not unethical. In fact, the processes are essential in major industrial applications especially food processing and flavoring.

Alcohol is pervasive in the food industry in its indispensable role as food soluble, flavoring and preservatives. These distinctive features of alcohol as solvent agents are also extensively applied in pharmaceutical, cosmetics, drugs and antibiotics, and other industrial applications. The total value of the application of alcohol in these industries is in the billions of dollars considering that just in the soft drink sector alone, for example, Coca Cola reported sales of USD22billion in its 2004 annual report (Azahari, 2010).

Alcohol occurs as a result of the processes of fermentation and in industrial applications, fermentation has proven to be an economically and commercially viable mode to produce alcohol. The type of alcohol that is commonly extracted through this process for the food industry is known as ethanol, which in its purest form would be harmful for consumption. Thus, ethanol is always mixed with other substances to render it safe for consumption. That brings us then to the main issue, is the prohibition on alcohol directed at alcohol per se or at the effects of alcohol, which is the intoxicating nature of alcohol.

In general there seem to be a lot of misunderstanding with regard to what alcohol actually is. The misunderstanding is due to the incorrect translation of term the khamr (Arabic) to alcohol (English). The term alcohol from a chemistry perspective is more than just ethanol. Alcohol is an organic compound in which the hydroxyl functional group (-OH) is bound to a carbon atom. There are many types of alcohol such as methanol, ethanol, propanol and butanol. Of those, ethanol is the type of alcohol commonly found in food, beverages, perfumes and medicines. Ethanol is derived from two main processes, hydration of ethylene and biological fermentation of carbohydrate source. Hydration of ethylene is the primary method for the industrial production of ethanol (synthetic alcohol), while fermentation is the primary method for production of beverage alcohol and fermented foods (Riaz, 1997).

Khamr is a term that derived from verb khamara means 'to shroud' or 'to cloud'. This term refers to any substance that intoxicates the mind and can cause the person who took it lose their ability to control their mind and action. The Maliki, Shafi'i, and Hanbali schools consider khamr to be any intoxicating drink made from grapes, dates, or raisins. Whereas according to Hadith Bukhari and Muslim, khamr is what covers intellect and made from five things that are grapes, dates, wheat, barley and honey.

The prohibition of khamr is clearly stated in the Holy Quraan. In Surah Al-Baqarah verse 219: "They ask Thee concerning Wine and Gambling, Say: In them is great sin, and some profit, for men; but the sin is greater than the profit." and also in Surah Al-Maaidah verse 90: "O Ye who believe! Intoxicants and Gambling, Sacrificing to Stones, and (divination by) Arrows, are an abomination, of Satan's handiwork; Keep away from such, that Ye may prosper."

Ethanol is not necessarily khamr although the intoxicating substance in khamr is ethanol. Ethanol produced from the non industry is not najs. Local fermented products such as *tapai, budu, cencalok* and *belaca*n are permissible to consume they contain ethanol as they are not intoxicating. The from a religious perspective and hence the halal compliance of such products will depend on the intention and utilization of the product. The usage of ethanol from khamr industry or it's by products and derivatives even in small quantity in food products is haram. The Fatwa on alcohol usage in food, beverages, perfumes and medicines have been revised by the Fatwa Committee of the National Council for Malaysian Islamic Religious Affairs on July 14 to 16, 2011. Based on the briefings, presentations and explanations presented by the experts of Halal Products Research Institute, Universiti Putra Malaysia and taking into account of the results of the dialogue held in the Fatwa Committee of the National Council for Islamic Affairs Malaysia earlier, the Fatwa committee has agreed to decide as follows:

- All khamr contains alcohol. However, not all alcohol is khamr. Alcohol obtained from khamr making process is najs and haram.
- Alcohol obtained from non khamr industry is not najs, but not permissible to drink in its original form because it is poisonous and can kill.
- Soft drinks that are processed or made not with the intention to produce khamr and contain alcohol below the level of 1% (v/v) is permissible to be drank.
- Soft drinks which are made with the intention and the same way as the process of making khamr, whether it contains a lot or a little alcohol or distilled alcohol are haram.
- Foods or drinks containing natural alcohol such as fruits, nuts or grains and its juice, or alcohol produced as by-product during the manufacturing process of food or drink is not najs and permissible to be eaten or drink.
- Foods or drinks that contain flavoring or coloring materials containing alcohol for the purpose of stabilization is a permissible to be used if the alcohol is not produced from the khamr source and the quantity of alcohol in the final product is not intoxicating, and at the rate not exceeding 0.5% alcohol.
- Medicines and perfumes that contain alcohol (not from khamr source) that being used as a solvent is not najs.

3. MATERIALS AND METHOD

Samples tested are procured from hypermarket around Kuantan including energy drinks, carbonated drinks, other soft drinks, juices and cordial, fermented and cultured drinks, flavouring and colouring. 95 samples were tested. Gas Chromatography equipped with mass spectrometry as detector was used in the study for detection and quantification of the ethanol compounds present. The . J & W fused silica capillary column (J & W Scienti®c, Folsom, CA, USA) of 30 m x 0.32 mm coated with a 1.8 mm film of DB-624 stationary phase was installed in an Agilent 7890 GC equipped with a split/split less injector and a mass spectrometry detector. Injection was carried out in the fast mode using the Agilent G1888 Auto sampler. Using helium as a carrier gas, with a constant gas velocity of 45 cm s−1, a split ratio of 5:1, and an oven program from 40 ◦C (2 min), ramp at 50 ◦Cmin−1 to 200 ◦C and hold for 2 minutes.

4. RESULTS AND DISCUSSIONS

95 samples tested and 58 samples found to contained ethanol. Table 1 showed the range of concentration of the ethanol detected in the market products. Only selected brands of energy drinks contained ethanol the highest make up to 0.142%. Sample with the highest ethanol content is one of the orange flavouring. Orange flavour is oil derived from orange skins and does not dissolve in water but dissolved easily in ethanol. Hence, the presence of higher ethanol content in orange-flavoured flavouring is detected. Vinegars are products from fermentation. As ethanol is one of the products of natural fermentation, ethanol was found present in all the imported vinegar samples. For local brand white vinegars (3 brands), no ethanol was detected. Study screened 10 different carbonated drink products, including different brands of cola. No ethanol detected from the selected tested samples. Ethanol compounds present in juices and cordials with high sugar content and especially in canned fruit syrup. Ethanol can be natural product from fermentation in those high sugar products or added as solvent for flavour and colour. Sparkling juices found to contain 0.05% ethanol. For malt fermented beverages that claimed free-alcohol, results showed an agreement with the statement, no ethanol detected. For all 5 brands of soy bean milk tested, results showed all five samples contained ethanol, ranges from 0.003-0.013%. For cultured drinks, brand 1 contained 0.004% for all 3 flavours tested, whereas other 2 brands, no ethanol detected.

Table 1: Different categories of beverages and condiments and the range of concentration of ethanol detected.

Food Samples	Number of Sample Tested	Number of Sample contained Ethanol	Amount of Ethanol Detected (%)
Energy Drink	10	5	0.002 – 0.142
Carbonated Drink	10	0	Not Detected
Tea and Coffee	19	4	0.002 – 0.053
Soy Bean Milk	5	5	0.003 – 0.013
Juices and Cordial	15	10	0.001 – 0.122
Fermented and Cultured Drink	16	5	0.001 – 0.078
Sparkling Juices	5	5	0.003 – 0.049
Vinegars	10	7	0.08 – 0.18
Flavouring and Colouring	5	3	0.002 – 3.510
TOTAL	95	58	

5. CONCLUSIONS

Study found that 58 out of 95 market food products tested contained ethanol compounds. Ethanol compound present can be product natural fermentation or added to aid in process. GC/MS is a suitable to detect small amount of ethanol present in complex sample matrix such as foods and beverages.

REFERENCES

E-Fatwa. (2011). Alkohol Dalam Makanan, Minuman, Pewangi Dan Ubat Ubatan. Retrieved from http://www.e-fatwa.gov.my

González-Arjona, D., González-Gallero, V., Pablos, F., & Gustavo González, A. (1999). Authentication and differentiation of Irish whiskeys by higher-alcohol congener analysis. *Analytica Chimica Acta*, 381(2-3), 257-264

Lachenmeier, D.W.; Rehm, J.; Gmel, G. Surrogate alcohol: what do we know and where do we go? *Alcohol. Clin. Exp. Res.* 2007, 31, 1613-1624.

Law, S. V., Abu Bakar, F., Mat Hashim, D. and Abdul Hamid, A. (2011). Popular fermented foods and beverages in Southeast Asia. *International Food Research Journal 18*: 475–484.

Riaz M.N., Chaudry M.M. (2004). Halal Food Production, CRC Press: Boca Raton, FL

Riaz M.N., Chaudry M.M. (1997). Alcohol: The Myths and Realities. In. R. Hill (Ed.), Handbook of halal and haram products. New York: Publishing Centre of American Muslim Research and Information.

WHO. Global Status Report on Alcohol; World Health Organization: Geneva, Switzerland, 2004

YOUR KNOWLEDGE HAS VALUE

- We will publish your bachelor's and master's thesis, essays and papers

- Your own eBook and book - sold worldwide in all relevant shops

- Earn money with each sale

Upload your text at www.GRIN.com
and publish for free